The Tractor Story

The Tractor Story

Trevor Innes and Duncan Wherrett

The History Press

Published in the United Kingdom in 2011 by
The History Press
The Mill · Brimscombe Port · Stroud ·
Gloucestershire · GL5 2QG

Reprinted 2014, 2017

British Library Cataloguing in Publication Data
A catalogue record for this book is available from the British Library.

ISBN 978-0-7524-6198-4

Typesetting and origination by
The History Press
Printed and bound in China.

CONTENTS

Did you know?
The word tractor was taken from the Latin *trahere* 'to pull'.

Whilst many people would consider the steam traction engine to be the first tractor, traction engines were generally heavy, cumbersome and not suitable for crossing soft or heavy ground. As a result, their agricultural use drawing ploughs was very limited. They were used mainly for powering farm machinery, such as threshing machines, by means of a continuous leather belt driven by the flywheel.

Traction engines were also used in pairs, dragging a plough or similar implement by cable from one side of a field to another. Where soil conditions permitted, direct hauling of implements was sometimes done, especially in the United States.

The basic definition of a tractor is a vehicle specifically designed to deliver a high tractive effort (or torque) at slow speeds, for the purposes of hauling machinery used mainly in agriculture and construction.

Generally a tractor is used in agriculture, as the one machine can be used for a multitude of tasks, such as ploughing, pulling seed drills, harvesters to reap the crops, etc. The advent of the three-point linkage systems on the rear paved the way for turning the tractor into a truly universal workhorse able to use hundreds of different implements. The result has been that the tractor and its associated machinery have replaced millions of workers, as one man today can do what dozens of men could do 100 years ago.

What made the tractor possible was the invention of the internal combustion engine for which the first confirmed patent was given to Alphonse Beau de Rochas in 1861. This led to Nikolaus Otto building the first

The Fowler traction engine (1876) is believed to be one of the oldest ploughing engines in existence. It weighed in at over 20 tons, and needed most of its power just to move itself along the ground. Its main work was done in tandem with another ploughing engine using the winch under the body of the tractor to pull two huge scoops to make dams. It could also drive shearing plants from the belt pulley, which it continued to do for over 70 years.

commercial four-cycle combustion engines in 1862. In 1876, Otto improved the four-stroke engine, working with Gottlieb Daimler and Wilhelm Maybach.

With the advent of the automobile with its internal combustion engine in 1886, petroleum products became more readily available, thus enabling the commercial success of the stationary engine as well. A range of manufacturers

Did you know?
The first recorded use of the word tractor meaning 'an engine or vehicle for pulling wagons or ploughs' was in 1901, replacing the earlier term 'traction engine'.

The Clutterbuck was a typical stationary engine used in the 1900s. Mounted on a trolley, the engine would be towed around the farm by a horse. Implements, such as threshers, were driven by a belt from the engine.

started making stationary engines of many differing horse-powers for a multitude of uses in the agricultural industry.

With stationary engines mounted on trolleys, it was not going to be long before a rudimentary steering system was developed and the tractor as we know was born. In the early years, it is a wonder their development continued as they were so unreliable, crude and prone to breakage.

The Charter is often credited as being the first petrol-engined tractor. Made in 1889 by John Charter in Dakota, USA, it was a 20hp single-cylinder open crank petrol engine lashed on to a Rumely steam traction engine framework. Similar machines followed in the 1890s and some were made in very small numbers. Examples came from Petter in Yeovil and Hornsby in Grantham, England. More came from America with the Huber in 1898 and the Kinnard-Haines who each made more than twenty machines.

Did you know?

Charles W. Hart and Charles H. Parr are known for being the founders of the tractor industry and are credited with coining the word 'tractor'.

Did you know?
Pair of horses with a single furrow plough could plough 1 acre per day.

In 1901, Dan Albone, a cycle manufacturer in Biggleswade, England, made the Ivel. It was unique for its light compact design and was designed specifically for farm work. It was the first real break from the other cumbersome machines based around traction engines.

A lot of the earlier tractors represent amazing achievements in engineering. Many things we think of as modern inventions were already done when mechanisation was in its infancy. In many cases, although the basic concepts had been invented, they required improvements in technology and engineering to make them fully effective. With the Renard Daimler Road Train, for example, the front and rear wheels of the trucks all steered so that they all followed a similar line. Designed by Frenchman M. Renard, it had a 110hp 6 cylinder Daimler sleeve-valve engine, and pulled up to six trailers which all had their central wheels driven from a propeller shaft running through them. The tractors went to India, Canada, Australia, the US, South America and Europe.

Many of the early machines were still basic stationary engines on wheels, such as the International Mogul 10-20. It resulted from the earlier development of the Morton traction truck to which the International

A Renard Daimler (1908) made by Daimler of Coventry. It had only done six trips when it broke down and had to be towed to its destination by a team of 60 donkeys.

An International Mogul (1910) the engine of which was fitted to a frame and wheels with a rudimentary clutch, steering and cooling system. It had rear wheels 70" high and 20" wide, with one forward gear of 1.75mph and one reverse gear, and was capable of pulling two five furrow ploughs, under the right conditions.

A McDonald Imperial (1908), which had a twin cylinder 30bhp engine 6.25" bore and 8.25" stroke. It weighed 6 tons and in 1974 it out-pulled a 110hp John Deere tractor. It had foot operated band brake on a live rear axle and a differential which had a locking pin in the left side wheel. 'It keeps your boys on the farm because it reduces drudgery and makes farm work more interesting' was its sales pitch at the time.

Did you know?

John Deere started making ploughs in 1837. They made their first tractor in 1918 and are now the biggest tractor manufacturer in the world.

A Caldwell Vale from 1910. The 1912 models were fitted with solid rubber tyres and three-way tipping bodies. This one has sand wheels 5 feet in diameter and 16" wide. Normal operating speed was 5mph with a load of 25 tons. It could haul up to 45 tons.

Harvester Company added their horizontal single cylinder four-stroke petrol stationary engine. Built from 1909 to 1914, 2441 Mogul tractors were made in Milwaukee, Wisconsin. Developing 20hp at 240rpm, the Mogul approached what we now regard as the modern tractor.

The Caldwell Vale is a case of revolutionary tractor and engineering development. Compared to the crude International Mogul, the Caldwell Vale was probably one of the most technologically advanced tractors in the world at the time, designed and made in Australia, and a fleet of them was used in the construction of the Australian capital Canberra. It was so successful even when loaded, it was said that the only thing that would stop it was an inch of rain.

The machine came with four-wheel drive, power steering and limited slip differentials as standard, and it was to be decades for such features became standard on tractors. The engine had a square bore to stroke ratio, which was unheard of at the time. The four cylinders delivered 80hp at 800rpm and there was a twin ignition system. With a fuel consumption of 1 mpg, a 60 gallon fuel tank was necessary.

During this period no-one really knew what a tractor was, so many different formats were tried with varying success; it was an age of constant experimentation. The Bates Steel Mule, although a strange looking machine, could supposedly plough 10 acres/day, harvest 40 acres/day and mow 36 acres/day in Australian conditions. If the machine did not break down, it was giving a performance greater than a team of horses. It can be difficult to compare plough rates across countries. In Britain, ploughing would be to a depth of 6–8 feet and the soil was likely to be heavy and wet; in Australia and the US the ploughing depth would be about 3 feet in lighter dry soil.

The Jelbart was another machine that epitomised the original concept of the 'tractor', which was the placing of a stationary engine on a framework in order to pull a plough by means other than horses.It had a working speed of 1–10mph, plus a reverse gear. Most tractors did not have such a speed until the 1950s.

There raged a controversy that was to last well into the late 1930s as to which was the more efficient – the tractor or the horse. As well as the issue of efficiency, generations of farmers had used and loved their horses. There was network of support industries for horse ploughing and the new machinery caused the same rows and

(Below) The Big Lizzie (1915) was a monster tractor made in Australia in 1914/15, with a 60hp Blackstone single cylinder crude oil engine. The vehicle weighed in at 45 tons, with an engine of 8 tons. It was 34 feet long, 11 feet wide, 18 feet high, with two trailers each 30 feet long. There is a picture of it taken in 1916 with 899 bags of wheat, weighing a total of 96 imperial tons or 107 US tons. It had unique wheels to spread the huge load and prevent it from sinking into the sand; platforms made of iron were attached to the wheels with hinges and cables.

problems that many new technological advances cause.

It has been estimated that 33 per cent of all the crops grown went back to feeding the animals and during a drought the stock sometimes had to be moved hundreds of miles to get food. In times of drought animals sometimes had to live on sewerage farms in order to survive. The tractor equation began to work when costs were

Did you know?
Many of the earliest tractors would use nearly a gallon of fuel to plough 1 acre.

This Jelbart (1916) was made in Ballarat, Victoria, Australia. The two-stroke single cylinder engine produced 6-10bhp at 625rpm.

compared and the increased speed of operations allowed farming to be much more efficient, both in man hours and in enabling all the crop to be sold for market.

The other big early advance that has been the mainstay of tractor production for over a century was made by Rudolf Diesel, his first patent being issued in 1894. After almost being killed by his engine when it exploded, he operated his first successful engine in 1897 and was granted his patent for his 'internal combustion engine' in 1898. The engine was designed to run on peanut oil and Rudolf Diesel hoped farmers and small industries could be more independent from the increasingly powerful fuel industries.

The Bates Steel Mule was one of the more bizarre interpretations of the tractor. The 1916 Bates Steel Mule was made by the Bates Machine & Tractor Co. in Joilet, Illinois. It had a four cylinder 30hp engine with a single track drive giving 3mph. These early machines were full of experimental features, like the Jelbart. It had a stack of three wheels, the first being the steering wheel, the second the clutch and the third the throttle. With extensions fitted, the tractor could be controlled from the implement behind with the driver sitting on

In this Bates Steel Mule (1916) one can see the stack of three steering wheels.

Did you know?

The first mass-produced tractor was the Fordson introduced by Henry Ford in 1917.

The engine for the Holt Renown (1919) had four separate OHV cylinders, with exposed rockers. It developed 45hp and could pull 4500lbs on the drawbar. The Holt weighed 13,900lbs and the engine alone weighed 2,300lbs. The 2 forward gears gave it 3.5mph.

the attached implement. In this way, one person could drive the tractor and adjust the attachment at the some time. The piston had two diameters, the larger being at the bottom of the piston, which helped move the air in the crankcase through the transfer port on the return stroke. The end result was a completely erratic firing stroke. The open final drive made it difficult for the owner to hear himself think. To start the engine, it was primed on benzene, started on lighting kerosene and run on crude oil and even mutton fat. It was notorious for starting fires.

Strong claims were made by early tractor manufacturers. The makers of the Holt Renown claimed: 'A short cut to work, with a 45hp caterpillar, through swamps impassable to men and horses,' In fact it often got bogged down to the top of the tracks and had to be pulled out by teams of horses. It was also heavy on fuel, using 112 gallons to plough 130 acres, and ultimately the running cost was its downfall.

The Moline was quite revolutionary in that it had an electric starter, electric lights and a battery ignition. Other features were its turning brakes and differential lock. However, there was only one forward and one reverse gear. It had a 4 cylinder 22 belt hp engine with 12 drawbar hp. The 1800rpm revs were high for its time.

The International Titan was one of the most successful of the early tractors, over 78,000 being produced between 1915 and 1922 by the International Harvester Co., Chicago. It was simple, reliable and a much more practical machine than the Mogul it replaced. It had twin cylinders giving 20hp at 575rpm and forward speeds up to 2.75mph.

The Emmerson was one of those tractors where the horse brigade might have been correct. They were known locally as the 'Everlasting Bastard' due to them forever breaking down, with badly designed

A Moline Model D (1920) driving a chaff cutter making animal feed. Concrete was added to the new model in the wheels to stop the machine rolling over. The rear wheels could be removed and replaced by various implements such as ploughs and cultivators. The two levers in front of the steering wheel are for the clutch and accelerator.

An International Titan 10-20 (1920) fitted with self-steering gear, so once the first plough run had been done it would then guide the tractor. One of the local farmers would let it loose ploughing the field and follow along with a team of horses and a second plough. Needless to say he had many hair-raising moments.

The Renault Crawler (1920) was made in France, along the lines of a First World War tank. It had a 24hp engine running at 1000rpm. One unusual feature was a slanted radiator in front of the driver. The radiator was cooled by turbine-type blades cast into the flywheel, which pushed air through the radiator. One of its more quirky attributes was a three-way dipstick for the oil. If the French instructions were not understood, and the tap was turned the wrong way the oil would simply drain out without the operator realising it.

This Emmerson Brantingham (1920), made in Rockford, Illinois, had a 13.16 rated drawbar hp, 23.2 rated belt hp engine which ran at 900rpm running on kerosene. It had two forward speeds up to 2.77mph and it weighed 4400lbs.

transmissions and water pumps. They were also noted for tipping over sideways due to their high centre of gravity and narrow wheelbase. Reliability was a great problem with tractors during this period. Farmers were rightly concerned about how well a tractor would actually perform in the field under different conditions. Horses they knew, but much of the new machinery was an unknown quantity.

A Cletrac Model W (1922) on a Yorkshire Steam Wagon (1903). This Cletrac was in use for 45 years and some models were used in sawmills up until the 1960s as the belt pulley running at engine speed was a great benefit. The Yorkshire was originally used for carrying bales of wool.

Many of the power figures coming from the manufacturers were also of uncertain accuracy. In the United States, for farmers to have some idea of the reality of the horsepower figures claimed by manufacturers, a bill was passed by the Nebraska Legislature in 1919, setting up the Nebraska testing facility in 1920. The idea was to make tractor information more reliable for the state's farmers. It remained the benchmark until 1979, testing 1750 tractors.

The Cletrac tractor proved so successful it remained virtually unchanged until 1931. It was fitted with an OHV Weidley engine producing 20hp. There was one forward and one reverse gear, and it used nearly two gallons of fuel an hour. It was able to plough 6–7 acres per day with a five-furrow plough.

The Yorkshire Steam Wagon had a T-type boiler working at 180lbs steam pressure. It could carry 6 tons and pull 20 tons on two trailers behind. A drive of 30 miles would use 112lbs of coal and 300 gallons of water.

The Peterborough was another very unusual beast, although it looked conventional. Its engine was originally developed as a tank engine for the British Ministry of Defence in the First World War. It has a two-step bore, the larger with the piston rings, the smaller with the piston skirt, making for a very long piston. The thinking was to prevent oil from the engine mixing with the fuel from the combustion process to make a smokeless exhaust so the tank would be less easily spotted.

When Henry Ford decided to make tractors, there was already a Ford Tractor Company. Also Ford shareholders were not

Did you know?

Although Ford were making the Model F tractor in 1917, the Model T car was considered so cheap that many kits were sold to convert them into tractors.

The Peterborough L30.4, (1922) had a four cylinder OHV engine producing 18 drawbar hp and 30-35 belt hp at 900–1000rpm, claiming it would pull 3750lbs at 2mph.

A Detroit Fordson (1923). The machine shared many parts with the Model T Ford including vibrating coils which made them difficult to start. Many farmers could not believe they could get so hot in summer and still freeze in winter.

This Fordson Trackson (1923) had a worm drive differential and simple brake drums for steering. This was pretty ineffective, so the motor would almost stall on turning. Driving was quite difficult and the driver was not protected from the tracks.

The Cletrac Model F (1923) was the smallest of the Cletrac line, rated at 9–16 horsepower. It sold for $595 at the time, and weighed 1820lbs.

The driver of this Glasgow (1924) once hit a large pothole and was promptly thrown off. The tractor, careering out of control, hit a tree, and because of the spikes, drove itself up the tree until it stood on its end, before falling over.

A Fiat 703 (1924). The 703 had a belt pulley which could use the 3 speed forward and one reverse gears; the wheel drive could be disconnected to allow this to happen. The drawbar could be raised and lowered, as well as moved from side to side

This example of a McCormick Deering 10-20 (1924) proved to be very reliable, remaining in use for 25 years. The model was made for sixteen years between 1923 and 1939, with rubber tyres being fitted to later models. It had overhead valves on its 4 cylinder in-line engine, again setting the standard for the future.

McCORMICK-DEERING
10-20 - H.P.

An Allis Chalmers 18-30 (1925), of which 16,000 were made between 1919 and 1929. 1922 was the worst sales year for Allis-Chalmers due to the success of the Fordson and the post-war depression. Peak production was reached in 1928. The engine was an excellent unit and used for numerous other applications. With two forward speeds and one reverse, this tractor used 108 gallons of fuel to plough 120 acres.

in favour of tractor production, so Henry Ford set up a new company and called the machines 'Fordson'. Fordson set the trend in tractors for a lightweight standard tread tractor, producing some 850,000 in ten years until 1928. Production was based in Detroit in the US and Cork in Ireland. Ford ceased making tractors in America in 1928, concentrating on cars instead while Fordson production carried on in England.

The Detroit Fordson was rated at 18–20hp at 1000rpm, they could pull 2400lbs, whilst the tractor weighed in at 2700lbs. It has three forward gears and one reverse. The Fordson Crawler started as a standard Dearbourne made in Michigan. The track conversion was made by The Full Crawler Company of Milwaukee, Wisconsin. The kit provided a full track vehicle for use on steep or difficult ground and was especially popular with loggers. Around 88,000 Fordsons were converted to Tracksons.

Unusually the Cletrac F was adapted to pull horse-drawn equipment. Some 64 years later in 1987 Caterpillar brought out their revolutionary new drive system on the new D7, which helped keep the drive out of mud. They were a little fed up when the owner of this machine pointed out to them that it had already been done so many years before. It was made by the Cleveland Tractor Company in Cleveland, Ohio.

The Glasgow, a three-wheel drive machine built by Wallace Ltd, in Cardonald, Glasgow, Scotland, was one of the weirdest tractors ever made. Not a successful machine, having failed in field trials against a 15–27hp Case, it remained unsold for several years until it was bought for belt work by one of Australia's 'Cattle Kings'.

The Benz Sendling S6 (1926) was prone to toppling over; the makers soon realised this was not the way to go.

The Benz Sendling seen here ploughing against a team of eight Clydesdale horses. Only 1188 were ever made.

The Hart Parr 28-50 (1926) was generally easy to start; put a little benzine into the priming valves in the cylinders, close off, pull it over to compression on one cylinder, give a good pull and away it went.

Did you know?

The world's first commercially produced fuel injection diesel tractor was the Benz Sendling made in 1922.

Car manufacturers were also getting into tractors. Fiat started making tractors in 1919, and its 703 had many unusual features, such as a planetary gear final drive built into the wheel hubs, a non-detachable cylinder head, and a clutch which you pressed down to engage rather than release. It had a 4 cylinder side-

A Hart Parr 12-24 Model E (1927), made in Charles City, Iowa, working a saw bench. It was known for its frugal fuel economy, using just 1.5 to 2 gallons per hour, its easy starting and reliability, having an efficient fan for cooling, and its drip feed total-loss oiling system from the Madison Kipp force-fed lubricator. After lubricating the engine, the engine oil was then channelled through pipes to the final drive gears – a total loss system.

valve engine producing 30 belt hp and 19 drawbar hp at 900rpm.

By the mid-1920s, many unsuitable designs were being dropped by manufacturers, while some design concepts were starting to be established. A forerunner of almost all future tractor designs, the McCormick Deering of 1924, had the

A John Deere Model D 15-27 (1926), seen here with a hand wire tie baler. Three men could make 400 bales a day.

engine, gearbox and differential all fitted as one built-up unit into a one-piece cast frame from the radiator to drawbar. It was one of the first machines to be fitted with a Power Take-Off to drive machinery. Formed from the great rivalries of two large grain harvesting companies, William Deering and McCormick, the International Harvester

This Lawson S6 (1926) was made in New Holstein, Wisconsin. It had a 4 cylinder Wisconsin motor producing 15 drawbar hp and 32 belt hp. There were two forward gears and one reverse.

Co. was established on 12 August 1902. There was talk of a merger nearly 20 years before, but the fierce competition had to take its toll before the merger was agreed.

The Benz Sendling was one of the biggest contradictions ever made. It had the world's first diesel engine, a marvel in its time, but a completely crude chassis and drive. The single rear wheel meant there was no need for a differential. The poor performance of tractors like the Benz continued to give encouragement to supporters of the horse-

The engine of the Lanz Bulldog (1926) had hot-bulb ignition and was started by putting a blow lamp under the bulb in the head to create a hot spot, which ignited the fuel. The engine was cranked over by a starting handle, which also doubled as the steering wheel.

drawn plough. The controversy continued well into the 1930s.

Another way of increasing the range and size for a tractor manufacturer was to do what Hart Parr did, simply put two 12-24 engines together on a common crankcase. It also had a dry sump, with drip feed total loss oil system from the Madison Kipp force feed lubricators – carbon dioxide emissions were not really frowned upon back then.

John Deere, founded in 1837 as a plough manufacturer that had expanded into other farm equipment, started making their own tractors in 1923 with the Model D, its twin cylinder engine generating 25.8 belt hp. It was produced from 1 March 1 1923 to 3 July 1953, giving it the longest production span of all two-cylinder John Deere tractors.

The Twin City 17-28 (1927) was uprated from the 12-20 of 1925 by increasing engine revolution from 1000 to 1075, to produce 17 drawbar and 28 belt hp. It was very popular and some 20,000 were produced.

Did you know?

Around 50 per cent of farming deaths have been caused by tractor rollovers.

The Caterpillar 20 (1927) was made in Peoria, Illinois. It weighed 7822lbs and sold for $1900. The OHV engine produced 20 rated drawbar hp and 26 rated belt hp at 1100 rpm. It had three forward gears giving a maximum of nearly 5mph, and a reverse gear.

A Holt 2 Ton Crawler (1927) produced by the Caterpillar Tractor Co. The Holt 2 Ton was one of the most popular tractors of the 1920s, being very reliable with an engine giving a long service life. It had a modern gear-driven overhead camshaft and fan developing 30 belt hp and 15 drawbar hp.

2 TON

A Caterpillar 30 (1928). Made between 1925 and 1932, some 23,830 were built. It had a four cylinder OHV petrol engine giving 34 belt hp and could pull 6120lbs.

A Cross Engine Case 12-20 (1927); the model first appeared in 1922 and remained in production until 1928, producing 13 drawbar hp and 21 belt hp with two forward gears of 2 and 3mph. It was a cast frame tractor weighing 4450lbs, yet was still able to pull 3150lbs. Its reliability was such that this particular example was working in a wood yard in 1971, having been in use for over 44 years.

A British Wallis (1927). Ruston and Hornsby in Lincoln in England made the British Wallis tractors under licence from the J.I. Case Plow Works, Racine, Wisconsin. They featured the same 'unit-frame' chassis as the US-built Wallis. Ruston & Hornsby chose to use their own four-cylinder 28hp vertical engine, which ran on both petrol and paraffin. The British Wallis took first place in its class at the Lincoln Tractor Trials in 1920, and Ruston & Hornsby produced it over the next ten years, during which time many were shipped to Australia and New Zealand.

A McCormick Deering 15-30 (1928) using its belt pulley to drive a corn crusher to make animal feed and bread flour.

A Bates Steel Mule F18-25 (1929). The Model F was in production from 1921 to 1937. Being a half-track machine the steering was by the wheels at the front instead of clutches or brakes as on full-track crawler tractors. It weighed 4850lbs, and its belt pulley was unusually mounted on top of the gearbox.

Lawson were famous for their quality products which enabled them to survive against much bigger companies for many years, but the Depression and several crop failures led to bankruptcy in 1935. They produced tractors from 1915 to 1935; one of their notable design features was the automobile-like driving position.

Built in Mannheim, Germany, by Heinrich Lanz, the Lanz Bulldog had a single cylinder two-stroke flap valve engine developing 30hp, and running on crude oil. Nearly 250,000 Bulldogs were built over 30 years. The earlier HR2s had a problem with their hopper cooling, but with the advent of the HR4 with an efficient radiator and fan, they became one of the most successful European tractors produced until the mid-1950s.

The Lanz Bulldog had a very unusual method of obtaining reverse gear – you simply ran the engine in reverse. It led to many dangerous situations, because if the engine idled too slowly or under an extreme load, it could suddenly reverse itself and run over the implement behind.

One of the classic sayings associated with a car burning excess oil and pulling into a petrol station was 'Can you fill up the oil and check the petrol please?' This probably came from the early tractors. The Twin City during 39 hours of tractor testing used 4.25 gallons of oil, 3 in the engine and 1.25 in the transmission. It wouldn't quite pass the latest pollution legislation.

The Caterpillar Tractor Co. was at the forefront of diesel tractor design which brought to light a lot of flaws. When it had roughly 1000 diesel-engine tractors in the field, problems with fuels and lubricants became evident. Although the company

THE BATES STEEL MULE

thought that the diesels would run on any grade of fuel oil, it was found that sulphur and waxes in the fuels were detrimental to engine life. This included stuck piston rings, scored cylinder walls and burned main bearings. It was discovered that oils with paraffin bases instead of asphalt bases did not cause as much sticking of the rings, so the company got together with Standard Oil of California to develop the first detergent oils. However, the company could not get the oil industry as a whole to adopt and distribute the detergent oils throughout the US, so many Caterpillar tractor owners could not get the new fuel. Caterpillar then enlisted its dealership network to distribute the necessary oils themselves.

Caterpillar tractors became very popular in South Australia, used primarily for dam-sinking. Land was divided up into 1 square mile blocks or 640 acres, and if a water course was part of the land, a dam would be sunk and allowed to fill with water, providing a supply for the dry months. However, the dams were not always water-tight, and having large surface areas to their depth, evaporation in the hot summer months meant they were not always of much practical use. The Caterpillar tractor made dam-sinking much easier as they could cope with steep slopes and had far greater traction.

Of the many design features developed during this period, some were destined for a good future. The McCormick Deering 15-30 made history in 1921 when it arrived with its unit construction and one-piece cast iron frame. The 30 belt hp four cylinder OHV engine made it one of the larger tractors of its day. It was said at the time to replace between 8–10 horses; it could pull 4190lbs at 2.4mph, being fitted with three forward gears of 2, 3 and 4mph. It became one of the most popular tractors of the steel-wheeled farming era, with many converted to rubber tyres. A less popular tractor was the Bates Model F Steel Mule, produced with a Roi four cylinder side-valve engine developing 18 drawbar hp and 25 belt hp. It had a honeycomb radiator, a Kingston carburettor and a 10-gallon fuel capacity. The steering box was of poor design and had a tendency to wrench itself to full lock, which certainly helped to make life a little more exciting and cause a few broken wrists.

The development of tractors did not proceed in an orderly timeline, as a myriad of inventors and manufacturers worked separately, dropping failures and expanding on successful models, so there were many different designs on the market at any one time. There were a number of landmarks, however, which affected

A McCormick Deering W30 (1933), a model built in Chicago by the International Harvester Co., which produced 19 drawbar hp and 31 belt hp. The engine had four cylinders with overhead valves and ran on kerosene. Three forward gears up to 3mph and one reverse were provided. This W30 gave trouble free service over its 25-year working life.

Did you know?
When rubber tyres started replacing metal wheels from the late 1920s, 25 per cent more power could be transmitted to the ground.

An Allis Chalmers Model U (1933). The rubber-tyred model had 4 speeds up to 10mph, whereas the steel wheeled one only had 3. It was fitted with an Allis Chalmers 5.2 litre engine of 25.63 drawbar and 35.04 belt hp.

Caterpillar

A Lanz Bulldog Crawler (1936) pulling a seven-furrow plough. There were relatively few made, and the one pictured was built from the remains of two collected from 800 miles apart.

The Caterpillar Twenty Two (1934) was basically an up-rated Twenty. It ran on petrol, with about one horsepower less when running on distillate.

The Case Model C (1936) updated the old fashioned and unconventional cross-engined models, having an in-line four cylinder OHV petrol/kero engine. It produced 16 drawbar hp and 25 belt hp at 1100rpm. The Case C had 3 forward gears up to 4.5mph, and one reverse gear.

the whole industry. One such was the pneumatic rubber tyre, although its introduction was a gradual process.

In the late 1920s, farmers were attaching rubber truck tyres to their tractors. Once tyre manufacturers noticed this, they started to produce bespoke rubber tyres for tractors. In 1931 the Goodrich Company began selling a steel-rimmed rubber tyre made specifically for tractors, and Firestone followed in 1932. It transformed the ability of tractors to be driven on roads and over greater distances, because they no longer left gouges in the road surface. With the rubber tyre, traction was improved, there was more effective use of the tractor's power and less damage was done to the land. However, they were not suitable for all environments. In Australia, during the land clearances and after, wooden stakes often destroyed rubber tyres, making them an expensive luxury.

Even during the Depression in the 1930s, many steel-wheeled tractors were converted to rubber tyres. For many years manufacturers offered models with a choice of steel or rubber wheels. The Allis Chalmers Model U made its name as the first production farm tractor to be offered with low pressure rubber tyres as an option when introduced in 1929. Priced at $1050 more than 20,000 were made until 1952.

The Caterpillar Twenty Two was one of the first of the yellow and black Caterpillar tractors, which became the standard colours after the merger of the Best and Holt companies in 1925. To save on manpower before Power Take-Off and hydraulics were fitted, farmers would sit on the implements and drive the tractor through rope reins.

The Oliver Hart Parr (1937) was the first of the in-line four cylinder OHV engines running on petrol or kerosene. It was built in the Oliver Corporation's factory in Chicago. Power Take-Off was standard and it had three speed transmission and weighed 3800lbs. In 1929, four companies joined together to form the Oliver farm equipment company and this machine was the last to carry the Hart Parr name.

There would be an extension steering wheel to control the steering, with a system of pulleys to control the throttle and clutch.

Lanz Crawler tractors followed their own line of development and their metal tracks had advantages in some conditions. Their advent made thousands of acres of once inaccessible land available for agriculture, allowing hilly land in grazing country to be sown with oats and barley. The low centre of gravity meant there was greater surface area to create traction, although steep descents could make the radiator spill over. It featured a 44hp crude oil engine, running at 630rpm, a plate clutch and a two range six speed transmission.

A testimonial from the time read, 'I am very pleased with our Case C tractor. I like the independent brakes to make square corners when ploughing. I also like the open wheels because it is nearly impossible to fill the lugs. I have done nearly all the field work on our 160 acre farm, although I am only 13 years old. Marion Bergren, Red Oak, Iowa.' The Case C came in three versions: the CC for row crop work; the

CO designed for orchards; and the CT for industrial use. A Power Take-Off shaft was available on all C models, as an option.

Daniel Massey and Alanson Harris were pioneers of farm mechanisation in Canada, and in 1891 their respective companies merged to form the Massey-Harris Co. Initially they sold other manufacturers' machines such as Big Bull and Parrett. It was the Parrett tractors that were the first to carry the Massey-Harris name. In 1928, Massey-Harris bought the J.I. Case Plow Works Co. based in Racine, Wisconsin,

A Massey Harris Model 27 (1937). The '27' engine produced 26 drawbar hp, running on either kerosene or distillate. It had a Power Take-Off unit fitted, so machinery could be driven at a constant speed from the tractor. Also it had a separate water tank which fed water into the engine to stop pre-ignition and could run cooler under heavy loads.

Did you know?

Britain has a tremendous heritage in the development and production of tractors. Now in the twenty-first century, the only tractors made in Britain are the New Holland and JCB.

An HSCS Crawler (1937) that was used for over 15,000 hours in 39 years, with only one new set of rings and several valve grinds. The tractors were so reliable and therefore popular that they were made with little variation for 30 years.

A John Deere Model D (1937) made in Moline, Illinois, and fitted with a twin cylinder OHV motor. It produced 37hp at 900rpm, running on kerosene. Three forward gears were fitted, giving up to 5mph, plus a reverse. Over 40 years ago the creek pictured was flooded and a woman in labour on her way to hospital was taken across by this tractor, as it was the only vehicle that could ford the water.

McDonald Super Imperial Super Diesel (1938). The McDonald Model TWB was manufactured in 1938 by A.H. McDonald & Co., Ballarat, Australia. Originally designed as a single cylinder stationary engine, it was modified to fit a tractor frame with gears – very successfully, judging by the reliability reports. It was started by a blow lamp heating up a 'hot-bulb' to provide the initial ignition, and thereafter the combustion keeps the 'hot-bulb' glowing red.

A McCormick Deering FI4 Farmall (1938). The larger F20 version came in 1924. The smaller F12 (the forerunner of the F14) was first made in 1932 and developed 11 drawbar hp and 15 rated belt hp. It had three forward gears giving 3mph. Production of the updated F14 by the International Harvester Co. only lasted for one year in 1938/9.

A Fordson Model N (1938). The American version was called the 'Allround', with a tricycle arrangement and V twin front wheels. They were basically the same as preceding models but with higher compression and faster engine speed of 1200rpm. They were noted for being temperamental to start, although generally pretty reliable.

This Oliver 90 (1939) tucked away in shed, became home to a destitute chicken.

OLIVER

The Ferguson three-point linkage allowed a wide variety of implements to be attached to the back of the tractor. The driver could now quickly attach and control an implement on his own. Via a control lever, hydraulics were used to raise and lower the implements.

which gave it access to the Wallis range of tractors. In turn, this led to the Massey Harris 25 using the Wallis revolutionary U-frame design.

In 1842, Nathanial Clayton joined with Joseph Shuttleworth in Britain and started producing ship steam engines, but they soon turned to agricultural machinery. Due to rapid expansion, European production

A Ferguson TE20 (1948), a model referred to as the 'Little Grey Fergie', or just a 'Fergie'.

was started in Vienna and Budapest, under Mathias Hofherr, who then united with Mr Schrantz. The group combined to form the Hofherr-Schrantz-Clayton-Schuttleworth organisation. They proceeded to make a complete range of agricultural machinery for world markets. The HSCS Crawler from Budapest had a single cylinder crude oil two-stroke steel flap valve engine of 7.5" bore and 7.5" stroke, giving 25 belt hp and 20 drawbar hp. Ignition was by hot-bulb and it was fitted with three forward gears up to 2.75mph and a reverse.

Made from 1924 to 1963, the Farmall range from the International Harvester Co. were the bestselling row crop tractors, being suitable for orchards and market gardens. Their narrow wheels set far apart gave the tractors a spindly and precarious appearance. Their tricycle design meant that they were not stable on hilly terrain and there were many deaths from tractor rollovers.

The Fordson Model N was made from 1937 until 1945. They were immensely popular and by August 1943 number

Harry Ferguson was born in County Down in 1884. His machinery is an integral part of tractor and agricultural history.

The power unit of this Ransome MG2 (1947) was a 6hp side-valve engine from the Sturmey Archer Cycle Gear Company. A centrifugal clutch engaged the single forward and reverse gears.

This John Deere Row Crop (1947) had a twin cylinder four-stroke forward facing flat engine of 20.6 drawbar hp and 23.5 belt hp. Remaining in production from 1935 to 1952, it was fitted with six forward gears giving 10mph and 1 reverse gear. Several different models were produced; for example, the 'BO' was for orchard work and the 'BN' had a single front wheel.

100,000 had rolled off the production line in Dagenham, Essex. Being quite a simple tractor it was ideally suited to the 'Land Army' when women were drafted into farming to replace the men being sent to the front lines. Having been painted orange or 'Harvest Gold', in 1939 the colour was changed to green; not quite as obvious to enemy bombers.

In its time, the Oliver 90 was without doubt one of the very best American

A Minneapolis-Moline Twin City UTS, (1947) featuring the streamlined hood, electric lights and fender of the new model line-up of 1939.

Did you know?

Seven Fergusons crossed the Antarctic with Sir Edmund Hillary in 1955–59. Some were half-tracks with skis; some had an extra wheel and were fitted with caterpillar tracks.

tractors. It first appeared in 1937 and remained in production until 1953 when it was replaced by the technically similar Model 99, itself produced until 1957. Power for the Oliver 90 was provided by an advanced four cylinder overhead valve 443cu. in. engine, using petrol, kerosene or distillate. It was claimed that the Oliver 80 was able to replace a 16-mule team and pull a four furrow plough.

A Fordson Major Half Track (1948) pulling a contour plough, shaping steep land to conserve water and prevent excess water run off and erosion. The land is ploughed to create a bank to trap the water, and the land is then worked between the contour banks.

The Model 90's advanced features included an electric starter, steel cylinder liners and an adjustable governor.

An implement such as a plough was attached to a tractor by a simple link – until Harry Ferguson came along. The Ferguson tractor has a unique place in tractor history. Harry Ferguson first patented the 2 hitch point draft control system in 1925. He added a third point in 1928, creating what we now know as the three-point linkage (hitch) system. Initially it had two top links and one bottom link, but he soon realised the problems and turned it upside down, with one link on the top and two on the bottom.

Prior to the Ferguson linkage, if the implement struck an obstruction, the link might break or the tractor turn over. Now an obstruction would result in the hydraulics lifting the implement and passing over it. Such was its success that nearly every tractor now uses a similar system.

The TE-20 (Tractor England) was produced in England from 1946 to 1951. It generated 20 drawbar hp and 25 belt hp. By

The Fordson Major Half Track (1948) had a side-valve four cylinder engine which gave 32.5 belt hp and using 1.5–2.5 gallons of kerosene an hour.

A Fordson E27N Major (1948). Perkins P4 and P6 diesel engines could be fitted to the Major and proved to be a successful after-sale conversion.

1956, approximately 1 million Fergusons had been made. The TO-20 (Tractor Overseas) came out in 1948 as an American version of the TE-20 and was a direct competitor to the Ford 9N. Ferguson made seven models between 1948 and 1960.

The range of tractors was expanding to meet the requirements of all types of agriculture. In the same way that small rows of cultivation required a specialist machine, so did market gardens and small holdings. A number of firms developed machines to meet this demand. Ransomes, Sims and Jefferies of Ipswich were a well-established company making engines, boilers and ploughs. They introduced the MG2 market garden miniature crawler in 1936. It was an immediate success. The MG2 was made until 1948 and with improvements, 15,000 were made over a 30-year period.

The John Deere Crawler Model MC (1949) had a four-cycle two cylinder engine running on petrol. Fitted with four forward gears, it was popular for orchard work.

Even before the Second World War, four cylinders in-line was becoming very much the standard engine configuration. After 1946 there was also increasing standardisation in the appearance of tractors; gone were the bizarre-looking experiments of the early days.

This Farmall A (1948) had a four cylinder petrol engine producing 13hp. With a four speed gearbox, the top speed was 10mph, and it could pull a load of 2387lbs at 2mph. Electric starting and lighting were fitted.

The Twin City UTS produced by Minneapolis-Moline (a company formed by the merger of Minneapolis Steel & Machinery, Minneapolis Threshing Machine and Moline Plow in 1929) featured an in-line four cylinder OHV engine with 30 drawbar hp. Its fourth gear gave it a working speed of 6mph and 20mph in

This Allis Chalmers Model B (1948) was one of 120,783 made between 1938 and 1953 in the UK and the US.

fifth gear, which was very useful for countries such as North America and Australia with larger distances to travel between fields.

The Fordson Major was made in Dagenham from 1945 and was based on the Fordson N. It had the same engine and transmission but in a new casing which

1949 Lloyd's Mark I Turner V4 OHV diesel produced 25 drawbar hp and 33 belt hp at 1500rpm. It was used mainly for dam clearing.

allowed for a PTO and hydraulic arms. A higher speed top gear was an option. Front end loaders were available, and it ran on kerosene.

A conversion kit from Roadless Tractors was available for the Major to create the Fordson Half Track. Three forward gears were fitted, driving the half-track arrangement, which seems to have been designed back to front, because when put into reverse, the tracks could run off the drive wheels and bend the back idler axle. The basic tractor, however, was pretty reliable. The Fordson name was dropped in 1964 when Fordson of England was merged with Ford of America.

John Deere tractor production dates back to 1837 when John Deere, a trained blacksmith, built his first plough in Grand Detour, Illinois. By 1847, he was making 1000 ploughs a year and moved to Moline, Illinois, to take advantage of better transportation facilities. The company first started experimenting with tractors in 1912,

This Chamberlin 40K (1950) was in use for 28 years. The 40K had a two cylinder side-valve horizontally opposed engine with 9 forward and 3 reverse gears, giving speeds up to 18mph.

The Turner 4V95 (Yeoman of England) (1950) and came with four forward gears giving 16mph. It could to pull 4500lbs at 2mph.

A Brockhouse President (1950) with a four cylinder Morris 8 car engine of 918cc, rated at 27.6hp at 4400rpm.

A HSCS Steel Horse (1952) with a single cylinder crude oil, hot-bulb engine producing 32 drawbar hp and 40 belt hp at 760rpm. Starting was by a blow-lamp placed on the head.

Did you know?

Around 1782, James Watt calculated a draft horse could pull a value of 33,000 foot-pounds per minute. This became known as 1 horsepower. Prospective buyers of steam engines could compare this to the output of an engine.

with a motor plough. However, it was not until the buying of the Waterloo Gasoline Engine Co, in 1918 that John Deere became a proper tractor manufacturer with an already proven design.

As John Deere didn't make crawler tractors, from 1940 to 47 the basic model BO chassis was shipped to the Lindeman factories at Yakima, Washington. Mounted on crawler tracks, the John Deere Lindeman Crawler emerged. Finally, the first all-John Deere crawler, the MC, was produced at Dubuque from 1949 to 1952.

Deere produced their first diesel-powered tractor in 1949 – quite late compared with some manufacturers – the Model R with two cylinders and 51hp. The diesel engine was started by a secondary petrol engine.

The Farmall A – manufactured from 1938 to 1948, then replaced by the Super A – and

The Field Marshal Series 3A (1952) was based on the Lanz Bulldog. Originally started with a cigarette paper or a gun cartridge hit with a hammer, the 3A had options of electric start and lighting, as well as 'Adrolic' hydraulic power lift.

Did you know?

The first turbo-charged tractor was the Allis-Chalmers D-19 diesel in 1962.

A Caterpillar D2 (1953). The main engine was four cylinder, four cycle, compression ignition, producing 24 drawbar hp at 1525rpm.

A Cockshott (1953) shredding pea straw for use as mulch for city gardeners, still in use 55 years after it was bought. It was purchased because at the time it was the most advanced tractor, having live Power Take-Off (it did not stop when the clutch was operated).

This Deutz FM2417 (1953) was found with the fuel pump rusted up after children had filled the diesel tank with water. Since repair, it has proved to be a very reliable unit and was used for broad acre farming.

The KL Bulldog (1953) had two-stroke hot-bulb compression ignition engine, producing 41 drawbar hp, running on crude oil fuel. It had six forward gears giving 14mph and two reverse gears giving nearly 10mph and could run on 1.5 gallons of fuel an hour.

the smaller Farmall Cub proved immensely popular on small acreage farms, as well as on parks and large lawns where it was fitted with an under-slung rotary mower. The whole drive train was built to one side to allow for increased vision in row crop work so it was an ideal choice for market gardeners.

Allis Chalmers were always an innovative company and the Model B introduced in 1938 was the first tractor in the world to be sold fitted with pneumatic tyres without the option of steel wheels. It was very popular until the Ferguson TE-20 came out and essentially made it redundant. Its four cylinder engine produced 19.51 drawbar hp and 22.5 belt hp with a four speed transmission.

Also in the small crawler range, the Lloyd was not one of the world's most successful

tractors. It was cobbled together after the Second World War, when there was a tremendous shortage of tractors. Using a Turner marine diesel engine, coupled to a Ford truck gearbox and rear axle, Bren gun carrier tracks were fitted. As it had

Monarch MK3 (1953) fitted with a Ford engine and gearbox producing 17bhp running on petrol or kerosene, which cost £279.

only Ford rear brake drums operated by levers, its steering ability was limited. Also the tracks were deficient, because the pins through the track plates kept breaking; they were simply not heavy enough for long term continuous use.

After the Second World War, there was a particular shortage of heavier duty tractors. In 1947, Harry Chamberlain of the Australian Ball Bearing Co. received a government subsidy and set up Chamberlain industries in an old munitions factory. The Chamberlain proved a reliable tractor, though rather fuel hungry.

The Turner 4V95 was made by the Turner Manufacturing Co., in Wolverhampton and was sold as a general purpose medium tractor: 'If you can tell genuine quality when you see it, you will pick the Turner – electric starting and lighting and it handles a 10–12 horse team load.' The Turner and the Lloyd were both powered by the same V4 engine although the Turner was more reliable and came with a 540rpm PTO and a belt pulley running at 1050rpm.

The President was made by Brockhouse Engineering Southport Ltd in Lancashire. Primarily designed for orchard, market garden and odd-job use with an optional lighting kit, it was a modest and useful machine. It cost £1500 in 1950, with a rotary hoe cultivator included. A wide range of accessories was provided, from a mid-lift linkage system featuring a double-bar tool frame for hoeing and ridging, to a three-point linkage worked off a gear pump, which slipped into the side of the transmission and could lift 5cwt.

The HSCS Company produced the Steel Horse, one of the most successful European

The maximum speed of this Lanz Alldog (1954) was approximately 12mph. A large front-mounted cargo box was available.

Fahr KT10D, 1954 ell it's pulling a plough so it must be a tractor. Fahr in Germany made a series of 'walking tractors' as single furrow hand ploughs, this one having a 10hp Sachs two-stroke diesel water-cooled engine. It was made from 1951 to 58. They merged with Deutz in 1961 to become Deutz-Fahr.

FAHR KT10D

This Ferguson TO-35 (1955) was a successor to the TO-20 and came in grey and gold, coming with 30 drawbar hp and 33 belt hp. Later production models were painted red with the gold engine. Production was from 1954–1960.

tractors; by 1950 there were over 50,000 in use worldwide. The noise and vibration of the tractors proved almost beyond human endurance, an owner declaring that the noise made him so ill he swore he would set fire to it, although he resisted the temptation.

The Marshall name first started in the English town of Gainsborough in 1848. Marshall were noted for their steam engines, both portable and traction. Around 1908–11 they had built a few motor traction engines – traction engines fitted with either a petrol-start paraffin engine or a two cylinder oil engine. In the late 1920s, with American internal combustion engined tractors taking over the market, they decided to use the Lanz Bulldog as the inspiration for a new range of tractors, one being the Series 3A, which had the same 40 belt hp single cylinder two-stroke diesel engine. Like all single cylinders, power output was at very low revs – 750rpm. The gearbox was six speed.

The D2 was the beginning of a new line-up of diesel Caterpillars which were started by a small petrol engine. The maximum drawbar pull of 6,778lbs was considerably more than equivalent horsepower wheel tractors of the time. Five forward gears were provided and it could turn in a radius of 10 feet. When exceedingly cold, the steering clutches became solid and impossible to steer, so it was necessary to unhook the implements and drive in a straight line until everything had warmed up.

The Cockshott was made by Cockshutt Farm Equipment Ltd. in Brantford, Ontario, Canada, The tyres were large at 15x34 (normal size was 15x30). The powerful

51 belt hp engine and 6mph working gear proved a winning combination. The 6 cylinder Buda motor had the American Bosch fuel injection with a compression ratio of 14.3:1 and produced 46 drawbar hp at 1650rpm. Top speed was 12mph.

The Deutz was manufactured by Klockner-Humboldt-Deutz, Cologne, Germany. It featured a twin cylinder diesel compression ignition four-stroke engine, producing 35 belt hp at 1350rpm. It had a 5 speed transmission giving 12.4mph. Although quite an advanced tractor in many ways, it was crude in others as it was still necessary to remove the rocker cover and oil the tappets every 5 hours.

The KL Bulldog was as a copy of the Lanz Model N from Melbourne in Australia. Renowned for their vibration, many farmers can remember going home at night with their bodies still rocking to the motion of the Bulldog. As they were difficult to start, some farmers would leave them running all night to keep them hot. This would stop

Did you know?
David Brown Tractors owned Aston Martin cars from 1947–72 and introduced and developed the DB range as used in the James Bond films.

the carbon build-up which otherwise would flake off and cause sparks. John Deere bought out Lanz Bulldog in 1962, in order to obtain their patented oil pump. They then closed the factory.

Designed for small cultivation, the Monarch had hydraulic lift and could pull a single furrow plough. Singer bought the rights to manufacture OTA (Oak Tree Appliances) tractors in 1953, the Monarch being the four-wheeled version. Singer made 293 tractors before production was cancelled in 1956 when the Rootes group bought them out.

The Lanz Alldog Tool Carrier Tractor came from the Lanz Bulldog stable in Mannheim and was powered by an air-cooled 12hp single cylinder diesel engine. The engine, the five forward with one reverse gearbox and the final drive housing were all mounted at the rear of the open tool frame. Standard equipment fitted to the Alldog included hydraulic linkage fitted at both ends, a Power Take-Off shaft, rear mounted belt pulley and independent brakes.

In 1953, the Massey Harris company merged with Ferguson to form Massey-Harris-Ferguson, with the name simplified to Massey-Ferguson in 1957. David Brown started building tractors with Harry Ferguson in 1936 as Ferguson-Browns. From 1939 they were made as David Brown tractors. Unusually their Crop Master tractor had a dual seat split in the middle, with the steering wheel in the middle so it had to be driven with the wheel to one side – not the most ergonomic driving position.

The David Brown 2D small cultivator was aimed at precision market garden work. It

The David Brown Crop Master (1954) had a four cylinder OHV engine producing 17 rated drawbar hp. A two speed Power Take-Off and belt pulley were included. Four forward gears were fitted giving a maximum speed of 12mph, and two reverse gears. The machine ran on kerosene and could pull 3700lbs.

The David Brown 2D (1957) had a 1262cc 14.1hp air-cooled two cylinder diesel engine which was lightweight and rear mounted.

This Holder (1958), seen here in its native orchard environment, steadfastly refused to get bogged despite the wettest weather conditions on record.

was also used on larger farms as a specialist row crop machine.

Small certainly did not mean low specification. It was designed to use mid-mounted implements with a rear lift and PTO as an option. There were two front mounted lifts which could be operated separately. If fitted, all three were operated by compressed air.

With 4 forward gears reaching 7.5mph and 1 reverse gear, it could pull 2082lbs.

The Holder, made in Metzingen Wurtt, Germany by Gebruder Holder Maschinen Fabrik, was four wheel drive with articulated steering, Power Take-Off and three point linkage. The five forward speeds and three reverse allowed it to pull loads of 1160lbs on its 5.00x16 tyres. It was fitted with a 12 belt hp Sachs engine.

China has the second highest Gross Domestic Product in the world. There are rapidly expanding high-rise cities and high-speed bullet trains, yet at least 500 million people are actively working the land (excluding the old and young) with only an estimated 2 million tractors between them. Most farmers have largely been using the same methods for the last 2000 years.

Did you know?

The first patent for four wheel drive on a tractor was in 1907, taken out by Felix and Norman Caldwell of South Australia. The made the Caldwell Vale tractor.

A single cylinder paddy field plough, with its paddle wheels and rotary hoe, takes a fraction of the time to work land than it would with oxen or by hand. After the harvest the wheels can be changed to road wheels, a trailer is attached, and the family can be taken to market.

A Dong Feng Crawler and Shanghai 4WD tractor. The Shanghai has a 35.3kw at 2000rpm 4 cylinder diesel engine with 6 forward speeds, 2 reverse. It is seen here near Kumming, western China. This village had three 'big' tractors to aid the villagers who work the land. Irrigation water from the village storage tank is distributed by water pumps mounted on little 'walking tractors'. Even though they have this technology, the villagers' houses are made of cobb with earth floors and the toilet is communal.

In the western world mechanisation and tractor technology have replaced around 90 per cent of the labour required for agriculture. The social and economic consequences of more mechanisation in China cannot be predicted. At the moment, in many areas each person is allocated 750 square metres of land to live off. Families plant rice side by

In the main tractor dealer in Lhasa Tibet; tractors were of varying sizes. The one in front had the latest direct injection single cylinder diesel engine. The SH240 produces 24hp at 2300rpm with two wheel drive. Surrounding the tractors were stationary engines and attachments for the little 'walking tractor' 2WD engine units.

High up in the mountains, this older 20hp single cylinder machine is still under constant use; the steam is rising from its hopper cooling. The people it had brought to the mountain field, however, were still working the land by hand.

A Fiat 666 in the Seychelles (1983). The 666 had a roll-bar fitted as standard although the cabin has been removed from this example.

side, creating beautifully manicured paddy fields. With little mechanisation, in 2009 China produced three times more food than the United States.

Most tractors in China are under 20hp and from the tallest mountain pass to the lowest paddy field the 'putt, putt, putt' of the little single cylinder is a distinct sound heard for miles. Being made in different sizes the smaller ones are called 'walking tractors'. A large range of interchangeable attachments such as mini seed drills, water

Did you know?

On the earlier tractors, due to poor fuels and poor combustion, de-cokes of the combustion chamber were necessary. An unusual, inexpensive method was to take hard grained rice and pour it into the carburettor.

A TAFE Massey Ferguson (1992) in India. In 1961, Tractors and Farm Equipment Limited (TAFE) was formed to market and manufacture Massey Ferguson tractors and related farm equipment in India.

pumps and single furrow ploughs make it a truly universal machine. In recent years, China has lost 15 per cent of its best agricultural land to development, while more machinery will drastically change the look of the countryside.

Developing agricultural economies require basic tractors rather than hundreds of horsepower, preferring models such as the Fiat 666. The Fiat develops 67hp from its 4 cylinder engine at 2500rpm with 62 PTO hp worked through a 20 speed transmission and 8 speed reverse. With four wheel drive, three point linkage and modest power, it more than fits the requirements for a small farm.

India shows an exponential expansion of mechanisation into farming, with only 8500 tractors being in the country in 1951 to an annual production of 140,000 units by the late 1980s and 270,000 by the late 1990s. It has now overtaken the US as the world's largest tractor producer, with sixteen national and four multi-nationals producing tractors. Despite this, it is still estimated that less than 50 per cent of the land is worked by mechanised means.

Although many technological advances were made surprisingly early in the history of tractors, they needed refinements, improvements and the will to integrate them into the machines. Four wheel drive and power steering were around in 1910 but it was not until after 1960 that they began to be used more widely.

Although the mainstream tractor manufacturers were advancing with new technical

Did you know?

Satellite navigation means a tractor can be driven to an accuracy of less than one inch without a human at the controls.

A Massey Ferguson 35 (1963). The 35 came out in 1960–1965 and followed on from the Ferguson TO-35. It also had the Ferguson three-point linkage.

A Trusty Steed Tractor Mk 2 (1965) powered by a 500cc air-cooled 14hp industrial engine.

Did you know?

The Case Corporation started in 1847 making threshing machines, then steam engines. They are now one of the largest tractor manufacturers in the world.

A Nuffield 4/25 (1968) Built by Leyland.

A Ford 3000 (1973). Built between 1965 and 1975 this small-holding tractor had a three cylinder 47hp diesel with 35 drawbar hp and either 4 or 8 gears. It was made in factories in the US, England and Belgium.

FORD 3000

16V-747
BIG BUD

developments, little oddities still popped up, like the Trusty Seed Tractor shown on page 101. Made in Barnet, Hertfordshire, England, this small garden tractor used a Norton (motorcycle) engine. Suitable for market gardens, it could pull a range of implements. Around 300 were made.

Most automobile manufacturers have built tractors and many still do. The Nuffield 4/25 came from the Leyland cars group in Britain. It had a choice of two engines, a 25hp 1500cc four cylinder indirect injection diesel engine, or a 27hp 1600cc petrol engine developing 61ft/lbs torque with 9 speed, 3 reverse transmission through a 3 range gearbox.

Over recent decades, power seems to have become almost limitless. 150hp is going to be enough to do most tasks in many countries; the grain prairies in North America are going to need much more. Many manufacturers produce machines of over 500hp but the real boss is the Big Bud. Made from 1961–1992, nearly 90 per cent made are still running, as they were so reliable they were just rebuilt often with increased power, for

The Big Bud is the largest tractor in the world, boasting 1110hp from its intercooled V16 Detroit diesel of 1472 cubic inches (24 litre engine). It can pull an 80 foot wide cultivator at 8mph cultivating 1.2 acres per minute. This model from 1977 was rebuilt in 2010. It uses around 55 gallons of fuel to plough 80 acres and regularly ploughs 20–25000 acres a year.

A Zetor 4911 (1978) with a 3 cylinder diesel engine of 45hp, 5 forward and 1 reverse gear, and a 2 speed ratio box doubling the number of gears.

JOHN DEERE

around one third of the cost of buying a new one.

Production of Zetor tractors started in 1946 in Czechoslovakia. They were quick to use front wheel drive, four wheel drive and air-conditioned cabins. The use of common parts on different models and efficient production methods, kept the tractors relatively cheap, resulting in large sales around the world.

Double wheels came along in the 1980s for extra traction and with four wheel drive machines like the John Deere 8650 4WD could be used in a greater range

The 290hp John Deere 8650 4WD (1986) could plough over 700 acres in 24 hours, compared to the John Deere Model D (1927), that with its 16 drawbar horse power could plough around 30 acres per day.

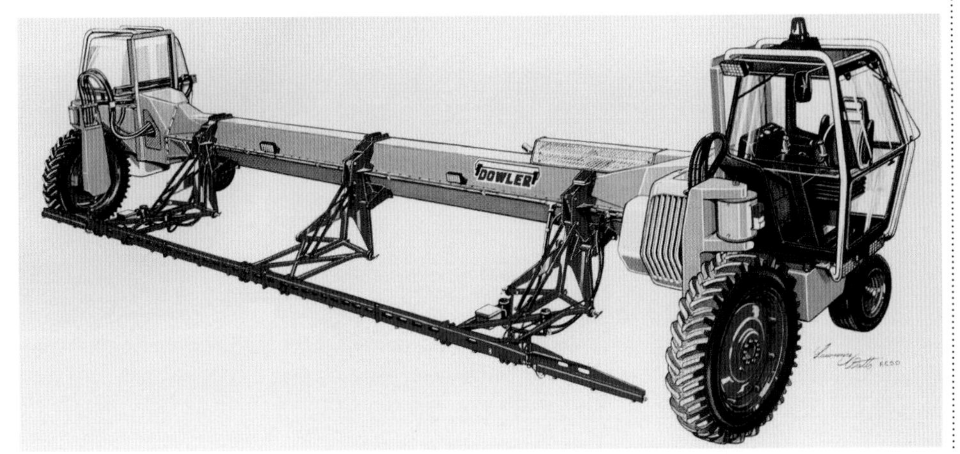

Several of these Dowler Gantry systems (1989) were made but David Dowler, the inventor, died suddenly, which ended this innovative tractor design.

A New Holland 7840 (1995), one of those made between 1991 and 1996. Early machines suffered reliability problems with the dual power transmissions. Dual power doubled the 8 forward gears and 2 reverse.

This John Deere Powerquad 6520 (2006) was made in Mannheim, Germany, it has a 4.5 litre turbocharged engine of 120hp at 2300rpm and 474nm of torque at 1500rpm. It developed 95 PTO hp. If you compare it with the earlier New Holland of similar size the power is achieved with an engine just two thirds the size.

of conditions. The growing power of the machines meant that they could do the work of 100 men in a fraction of the time.

One of the most innovative of all tractor designs, the Dowler Gantry system (see page 109) is designed around a central beam 12 metres long, to which you attach different implements for cultivation, seeding

CAT
Power

and spraying. Hydrostatically driven, each end can swivel through 90 degrees so that it can be driven on the road, being a long rather than a wide vehicle. The seat and controls swivel, allowing 360 degree vision.

New Hollands were first made in 1895 in Pennsylvania and became part of Ford in 1985. In 1990, Ford sold its farm machinery business to Fiat and in due course the Ford name was dropped.

Later New Hollands feature a more rounded and streamlined body. The New Holland 7840 featured a 6.6 litre non-turbocharged Ford Powerstar six cylinder 90hp engine. Improvements during its lifetime made it a more reliable machine with clutchless gear shifting, speed matching in gears and electronic draft control. Since being introduced in 1962, turbocharge has become a standard method of increasing power output without increasing engine size.

As rubber tyres dominated tractors, Caterpillar began to focus on construction equipment. The development of rubber tracks meant Caterpillar were able to produce farm tractors again in 1987 and they began to be sold under the Challenger name. Caterpillar sold the line to AGCO in 2002, although they are still sold through Caterpillar dealers.

The huge rubber tracks now make crawler-type tractors practical for use on the road, finally allowing them the advantages of speeds of 25mph.

The Massey Ferguson 7495 comes from the mighty AGCO corporation as does Fendt, hence it shares the same advanced engine and variable transmission technology but under the different name of Dyna -VT.

A Challenger MT865B (2008). Rated at 510hp but tested at 543 PTO hp and 477.2 drawbar hp, unusually the Challenger seemed to out-perform the makers specifications. The engine is a Caterpillar Diesel Type six cylinder vertical with turbocharger and air-to-air after-cooler. Compression ratio is 16.5 to 1 displacing 1105 cu in (18130cc). It has 16 forward and 4 reverse selective gear fixed ratios, with full range operator controlled power shift transmission making it one of the new breed of monster tractors. Electro-hydraulic differential steering is used.

Did you know?

Formula 1 racing cars have a clutchless gear change; flicking a lever changes gear. This came out on tractors back in 1964. It was on the Ford 3000 with 10 forward gears and 2 reverse.

A Massey Ferguson 7495 (2009). The 7495 is 6.6 litres producing up to 180hp with 836nm of torque. It has been tested against other manufacturers machine of similar horsepower and the fuel savings were 10–16 per cent.

It has a six cylinder SisuDiesel engine with the latest anti-pollution technology.

Flotation tyres have been another significant development in recent years as they are designed to run on very low pressures, such as 12 psi. The wide tyre and low ground pressure avoids ground compression and the wheel is less likely to become bogged.

Flotation tyres are now often used instead of dual rear wheels.

One of the biggest advances has been the rapid development of satellite navigation. Especially useful at night, crops can be sprayed with wide booms to an accuracy of 20mm. This allows more effective treatment as the plants have a better uptake of the chemicals with the dew at night, whereas during daytime the plants are more stressed. It is also possible to use the satellite for electronic driving of the tractors, again with the 20mm accuracy for ploughing,

CAUTION
CFX-750
OUTER FOLD
SPRAY BOOMS

Sat-nav is not a gimmick on a modern farm. Soil quality is mapped then just the right amount of fertiliser can be added to each section of soil.

A John Deere 7530 fitted with a 6.8 litre turbocharged engine with air-to-water after-cooling, the 7530 produces 180hp at 2100rpm and a 153 rated PTO hp. A fridge is an optional extra.

seeding and harvesting, cutting down on wastage of time, fuel and materials. The same tyre tracks (tramlines) can be used for all operations. With agricultural fuel prices rising 100 per cent in two years, the savings can be enormous.

For the user, one of the most significant advances is the air-conditioned cabin. When a driver is spending hours at the controls in summer heat, this is a great asset.

Fendt are often seen as the élite of tractors because of their build-quality and advanced features. Probably the most advanced tractor in the world in 2011, the Fendt has produced the world's first hydrostatic drive for high performance 260hp tractors. It has one pump and two hydraulic motors which give continuously variable gearing from 0–19mph and 0–31mph in its two gear ranges. The motor

A Deere 7530 rear linkage. Considerably more versatile and complicated, the rear end of the modern tractor still carries the tradition of the Ferguson three-point linkage.

The control panel of a Deere 7530 has a choice of three transmission types: PowrQuad Plus 16/16 forward and reverse, designed to shift between gears without clutching; AutoQuad Plus 20/20 and 24/24 with automatic shifting; and continuously variable transmission, which co-ordinates the electronically operated engine and transmission for fully automatic operation between 0 to 25mph.

The Fendt 415 Vario (2009) produces 155hp, and has 3 PTO speeds and is available with front PTO as well. It has front suspension, cab and linkage suspension systems to allow a smooth ride and reduce driver fatigue.

is electronically controlled to maintain a constant drive speed and if the load drops the engine revs by more than 10 per cent, it automatically adjusts the ground speed to match the engine speed. Drivers can either dial in their required speed or let the tractor adjust to the load. By working at ultimate engine and transmission efficiency, Fendt claim that 16 per cent saving can be made on costs of fuel and labour, resulting in the average American farmer saving around $7500 per year.

Other manufacturers have introduced their own versions of continuously variable transmission which all allow the transmission to move from the lowest gear ratio to the highest in a stepless and smooth manner. It is not restricted to fixed gear ratios. The drive shaft can maintain a constant velocity whatever the

This Fendt 933 (2011) is capable of running on rape methyl ester as well as diesel, with no engine modification. The 7.4 litre 4 valve engine develops 333hp.

Did you know?

The largest tractors in the world can plough an acre in less than a minute.

output velocity. The engine can run at its most efficient revs for any vehicle speed. Alternatively, the engine can use the revs which will give peak power.

A top of the range Fendt represents the peak of tractor technology for 2011. It has a two-range continuously variable transmission capable of 37mph forward and 18.5mph reverse. The 933 has 300mm of hydropneumatic independent wheel suspension, and fully electronic engine and transmission management control, by a joystick. It is fitted with Auto-GuidePro which is the automatic steering and satellite navigation system that reduces fatigue levels and provides maximum land coverage with minimum waste.

It remains to be seen whether the new electronic technology will stand the extremes of climate, plagues of mice and other agricultural hazards that so many of the earlier tractors have weathered.

1712 The first commercially successful steam engine.

1769 James Watt granted a patent for his improved design of the steam engine.

1837 John Deere, a blacksmith, started making ploughs.

1842 The first steam-powered traction engine was probably made by Ransomes of Ipswich, England.

1847 The Case Corporation is founded by Jerome Increase Case and sells threshing machines. Case began producing steam engines in 1869 and in 1878 introduced its first steam traction engine.

1847 The McCormick brothers start making a horse-drawn reaper. They were to become International Harvester.

1847 Allis Chalmers starts as a maker of small buhrstone and begins making steam power machinery 22 years later.

1861 The first patent for the internal combustion engine is given to Alphonse Beau de Rochas.

1862 Nikolaus Otto buildst the first commercial four-cycle combustion engines.

1889 The Charter is the first successful tractor using a petrol engine, with a Rumley steam traction engine chassis.

1891 Massey-Harris formed with the merger of the Massey Manufacturing Co. and A. Harris & Sons, who were major manufacturers of harvesting equipment in Canada.

1894 Rudolf Diesel obtains his first patent for the diesel engine, designed to run on peanut oil.

1901 Charles Hart and Charles Parr in the USA build the first factory for making a traction engine powered by an internal combustion engine tractor. Hart and Parr are credited with first using the term 'tractor' for the traction engine.

1902 The Ivel is the first light weight petrol powered tractor specially designed for general farm use. It was built by Dan Albone, a cycle manufacturer in Biggleswade, England.

1902 The International Harvester Company is formed by the merger of the McCormick Harvesting Machine Company and the Deering Harvester Company.

1904 The first successful crawler tractor to be fitted with tracks rather than wheels. Made by Benjamin Holt, a manufacturer of agricultural equipment in California. It is powered by steam with a petrol engine version in 1906.

1906 The Gourgis from France is the first machine to use an implement powered directly by the engine – the forerunner to Power Take-off.

1907 Henry Ford makes his first experimental petrol engine tractor.

1907 The first patent for four wheel drive on a tractor was taken out by Felix and Norman Caldwell of South Australia. They made the Caldwell Vale tractor.

1914 Allis-Chalmers makes its first tractor, the 10-18.

1917 The first mass-produced tractor is the Fordson introduced by Henry Ford. It soon made up 50% of the worldwide tractor sales.

1918 John Deere start selling Waterloo Boy tractors after buying the company.

1921 The International Harvester 15-30 is the first vehicle in full production with three methods of delivering power: power from traction, belt pulley and Power Take-off.

1921 The successful Lanz Bulldog single cylinder two stroke is launched. It could run on waste oil. Made in Germany, it set the standard in its class until the 1950s.

1922 The Benz Sending is the world's first commercially produced fuel injection diesel tractor.

1923 The Model D is the first tractor with the John Deere name.

1924 Farmall introduces the first successful rowcrop tractor.

1924 The McCormick Deering has the engine, gearbox and differential all fitted as a single unit into a one-piece cast frame from the radiator to drawbar. This unit construction is the forerunner of most future tractor design.

1925 The Caterpillar Tractor Company is formed with a merger of the Holt Manufacturing Company and the C. L. Best Gas Traction Company.

1925 Harry Ferguson patents the 2 hitch point draft control system used with a plough on Fordson tractors.

1928 Ferguson makes the 3 point linkage system. With hydaulics, it could raise and lower attached implements like ploughs, cultivators and mowers, and automatically set them to the required depth.

1931 Caterpillar manufacture a crawler tractor with a diesel engine giving more power, reliability, and fuel efficiency than petrol engines.

1932 First pneumatic rubber tyres specially made for tractors.

1933 The Allis-Chalmers Model U is the first production tractor offering pneumatic rubber tyres as standard.

1933 Fordson production starts in Dagenham, England.

1933 Three point linkage system first appears as a prototype on the Ferguson Black.

1934 John Deere introduces a hydraulic lift on its rowcrop tractors.

1935 McComick-Deering WD-40 became the first diesel-powered tractor manufactured in the USA.

1936 First commercial production of the 3 point linkage with the Ferguson Model A made by David Brown.

1937 Allis-Chalmers Model B is a pioneer of the new small lightweight and inexpensive tractors, which could plough, cultivate and harvest.

1939 Ford start making tractors with the Ferguson 3 point linkage system.

1946 Ferguson produces the TE-20. The first tractor under the Ferguson name, it was known as the 'little grey Fergie'.

1947 All major American manufacturers started moving over to multi-cylinder diesel engines.

1948 Cockshutt 30 is the first production tractor with live Power Take-off. Normally, PTO would stop when the clutch was pressed. Two clutches, one for the engine, one for the PTO, avoided this.

1960 John Deere are the first to begin large production of four wheel drive tractors with the 8010.

1962 The first turbo-charged tractor is the Allis-Chalmers D-19 diesel.

1964 The Ford 3000 was the first tractor with a clutchless gear-change with 10 forward gears and 2 reverse.

1978 Eight wheel tractors for better traction introduced by the larger firms like Deere, Massey-Ferguson and Big Bud.

1994 Global Positioning System (GPS) receivers used to record precise mapping of soil conditions and control the exact quantities of fertiliser, pesticides and water required.

1995 Fendt produce the first tractor with Continuously Variable Transmission giving step-less transmission with no gear changing, enabling the machine to run at optimum engine and power efficiency.

Recent improvements in tractors concentrate more on increasing power, more precise satellite navigation, greater cabin comfort and engineering refinements.